LUWA ADEBANJO

A Visitor Who Belongs Here

First published by Luwa Adebanjo 2021

First edition

Cover art by Bolusefe Akande

This book was professionally typeset on Reedsy.
Find out more at reedsy.com

This is for every survivor who is struggling to keep going.
Don't get stuck on this chapter, there are better pages ahead.
Your celebration is coming.

come celebrate
with me that everyday
something has tried to kill me
and has failed.

'WON'T YOU CELEBRATE WITH ME' BY LUCILLE CLIFTON

Contents

II A little while later: The Unmaking of a Poet

III After OCD: Hunger

Foreword

There are many ways to tell a story, and when I started this anthology I thought I would be telling a sad one. In my head, my life was all about struggling to fit in and never quite making it, and it made me feel bitter and angry. I didn't think to look at the story of my life with a different lens until I read a poem by the incredible Lucile Cliffton, titled: *won't you celebrate with me.* It is a beautiful poem and the first one that made me feel utterly bare and completely understood. In school, the poems we studied were historically relevant accounts or romantic sonnets. We rarely read anything by a black poet and nothing by a black woman. The representation I found in Lucile Cliffton's work genuinely changed my life.

I had always felt that my story was marked with trauma, and I had never looked at it another way.

In reality, for every bit of sadness in my life, I needed equal amounts of strength to fight back. I never subscribed to the idea that we had to suffer to learn or grow- so much suffering is pointless and gets no consequence- but I had also not focused on how hard it is to get through suffering. When I looked back on my life the way Lucile had I realised that my story was one of celebration, not sadness. I couldn't look back with anything but wonder. *How on Earth did I get through my childhood?* How did I make it here? I was *so proud.* Younger me didn't have a diagnosis but she still had pain, and now I had the help needed to heal that pain. Writing this anthology forced me to look back and confront my past, and it made me realise I wanted to celebrate. I assumed that I wouldn't be able to see beyond my trauma, but I could.

I survived, and that meant something. I often experience enouement, which is the feeling you get in the present when you cannot fulfil the urge to go back and tell your past self that it's all going to be okay. This anthology is an attempt to deal with my feelings of enouement, I can't go back and tell my younger self there is a reason to keep going, but by living 13-year-old-me's dream I get as close as I can. However, that was only possible when I realised that my story belonged to me alone.

When I began to take the time to be thankful for the people, skills, lessons and tools that helped me through my worst times I realised I had to change my anthology. I am not the same girl I used to be. I have a family, good friends, a support network, work I am proud of and goals I plan to reach. I deconstructed my anthology completely, going through the poems and showing a different story: one of hope.

Therefore, I encourage you to read the poems once and feel all the emotions inside of them. Connect to the sadness, loss, and anger. Consider similar moments in your life, remember your own struggles as much as you can comfortably and just react. Then, go through and read the anthology again, through the lens of celebration. Look at the growth in your own life, and the strength it has taken to keep going. The fact that you are here is proof that you are strong enough to keep going. I hope doing this will enable you to get as close as humanly possible to going through my life with me. Go once through these poems as I did, scared, alone and unheard. Then go again through celebration, and see the strength and power despite it all.

There are many ways to tell a story. This is mine, and it is a celebration.

I

In the Beginning: A Time for Many Words

As an immigrant coming to the UK at a young age I struggled to fit in, constantly feeling like a visitor wherever I went. My childhood was hard, and there were many times growing up that I wished I could run away from it all. When I looked through my childhood poems, the first thing I noticed is how much I wrote. The sheer volume of it was incredible.
Notebook, upon notebook covered in words. I had so much to say, and no one to say it to. There are so many poems that could be here, and all of them spoke to how anxious I was as a child, my mind constantly spinning itself into exhaustion. The added culture shock of now being Black made matters worse. When I was in Nigeria, I wasn't seen as a Black girl. I was just a little girl, it was only when I moved countries that I became a minority.
At school, instead of the 'funny' or 'bossy' one, I was just the Black one. At home, I was struggling with helping raise my younger siblings, I tried to convince myself that I could be the perfect daughter and it was taking a toll.
And in my head, I was questioning my faith and sexuality.

One

Grief

I wish to mourn that girl,

Mourn for her, I mean.
The girl who wasn't ~~molested~~.
The girl that never had an intrusive thought.
The girl whose loved ones loved her.
The girl who never flinched when someone raised a hand at her, for she only knew gentle touches.
The girl who did not know what it felt like to cry so much she would dry heave for what felt like an eternity.

I imagine her sometimes,
She only knows light and love and laughter,
She abhors violence because she has never seen evil,
She trusts readily, has no secrets worth spilling
She is soft and loving in all ways and in all things,
She is smart and funny and brave.
She is kind and fearless and confident; self-esteem rolls off her like waves.

She is not dark, not
broken.
She is whole and lovely.
She never doubts- she need not question every voice in her head.

She just is.

Except she isn't-
She never was.

And you cannot mourn someone who never got the chance to live.

Two

Go back home

The other day,

A man told me to

"Go back home!"

I wanted to tell him,

"I am home!"

But then I thought

There's no point

To this man, I am just an ungrateful visitor
my accent, my memories, my red passport,
none of that matters to him
to him, I don't belong here

maybe he's right
if he belongs here,
I don't want to.

Three

Exodus 20 King James Version (KJV)

them: I am the Lord

Your God.
Ye
shall
therefore
shall do my
judgments, and keep mine ordinances.
shall ye not do: neither shall ye walk in their ordinances.
After
the doings of
themselves
with mankind,
as with
womankind: it is
abomination.
Know ye not that
the land is

defiled
which I
cast out
before
you.
shall ye
keep my statutes, and
my judgments, and shall not inherit
the kingdom of God?
Be
not deceived: neither fornicators, 10 Nor thieves, nor covetous, nor drunkards, nor
revilers,
nor idolaters, nor idolaters,
nor idolaters, nor
idolaters, nor idolaters, nor idolaters,
nor idolaters, nor idolaters, nor idolaters, nor idolaters, nor idolaters,
nor idolaters, nor
idolaters, nor effeminate,
nor
abusers of
the Lord your God.
Lord your God is a sabbath day by
keep
my commit adultery.
Be not deceived:
neither shall commit
any
of
these abominations,
these things: for in all these things:
For this
cause God gave them

up unto
vile affections: forever. Amen.
For this cause, God gave them up unto vile affections: for
ever. Amen.
this cause
God gave them up
unto vile affections:
forever. Amen.
For this cause God gave them up unto vile affections: forever. Amen.For
this
cause God
gave them up

Four

She Likes Boys

She does, she really does. She's always liked them.

Ever since she was young, she loved to play with them: kiss chases across the playground.

She always knew, as *soon* as she saw them if they were going to be her *best friend.*

She sometimes gets obsessed with them. The way they looked, the way they smiled. She loved making them laugh, such a high, pretty sound.

She likes boys

She dreams about them sometimes, at night, in her bed. Waking up to bedsheets covered in sweat, dampness between her thighs.

They were confusing dreams because they weren't wearing much clothing.

She didn't care though, because they were gorgeous. And in these dreams sometimes she would get distracted, and instead of their faces, she would linger on the curve of their breasts, the softness of their stomachs, the thickness of their thighs...

She likes boys

She is waiting for one. A boy.

She is waiting for one, she is saving herself for him.

Her first kiss, her first caress, her first everything.

She dreams about the first time she will hold their soft hands in hers, and kiss their sweet, blush tinted lips.

She is waiting for one.

She is waiting for a boy.

She likes boys

She loves them.

She loves them with long, thick hair you can run your hands through. Or soft hair that is kinky, and tightly curled, that bounces back when you gently tease it.

But short hair is fine! Short hair is great! Anything in between is fine. She loves them all.

She loves them when they wear makeup and powder their cheeks rose pink or raspberry red. Soft matte lipstick, or sticky sweet lip gloss. They look good made-up or wearing nothing at all.

She loves when they wear dresses or long skirts that lift when they spin around.

She loves when they wear trousers, shorts, skinny jeans. *Oh, my god.* She *loves* skinny jeans.

She likes boys

She loves the plump softness of their lips. The shy, tentative strokes of their tongues. She loves their perfume; they often wear perfume. Flowery, smoky scents that linger on her pillowcases.

She likes boys

She loves to cook for them, meals that they share together, holding hands across the table. She loves cooking *with* them, blasting afrobeats and britpop as they dance around the kitchen.

She loves cuddling them in bed after they eat. The simple way they squish into place, legs entangled. She loves snuggling up to them, their face buried in her neck. Their hands under her shirt, stroking her full belly.

She loves being held and looked after. She loves whispering secrets to each other in the dark.

She

Likes

Boys

She loves dressing up for them. In playsuits, dresses, crop tops, boyfriend jeans, or lacy sets from Ann Summers.

She loves when they compliment her; calling her pet names that make her melt. *Princess, my love, darling, angel. Give me a kiss.*

She

Likes

Boys

She does. Because that's how it's meant to be.

You're meant to love boys

Boys love girls. Girls love boys.

That's what she follows, what she believes.

Anything else is wrong. No matter what she feels.

That's why

She likes boys

She was obsessed with the women's world cup. She could watch it for hours and she knows nothing about football. She used to say she hated it. But there was just something about the players she couldn't take her eyes off.

She likes

Boys

She likes the idea of marrying one, one day.

Could they even get married?

Sometimes she is unsure. She questions everything.

Maybe she could like something different.

But that, for now, is not meant to be. For now, **she likes boys**.

Five

Petition A for the will to live

Please tell me that one day I love me.

Please tell me it's all worth it.

Tell me anything at all. Tell me a beautiful lie. Weave it in-between my hopes and wishes. Coddle and caress me with a lovely truth.

Do whatever you have to.

Just convince me.

Please.

Six

Our love is Pink

Bright pink, bubbly, flirty and sweet,

Not pink because of romance or any other cliché
But pink for joy, flower petals and glitter,
It's bold, and it doesn't back down.
It's loud and argumentative
It's silly and whimsical
I was scared, but I was willing to trust you.
Willing to see where this goes, knowing wherever we end up it will be worth it.

I don't know if it was. I thought you would be my first love, but pink fades quickly. It is easily tarnished. Like a fine blush powder, we spread quickly but melted away fast.

In our triumph, we died
In our kiss, you consumed me so entirely
I don't think I'll ever be the same

Seven

Visitations, Vacations and Vacancies

They say,

"Wherever you go, there you are."

And I believe that is true.

I wish I could go somewhere without me.
Or at least without you.

Eight

mother/sister/father/daughter/brother/friend

Can you make me breakfast?

Can you take them to school?

Can you take me to the park?

Can you cook dinner?

Can you help them with homework?

What should I do about the bullies in school?

Take care of your brothers, daughter

Clean up the kitchen

Play in the garden with me!

I think I like someone, I don't know what to do, please be my friend

Hold my hand I'm scared
  Me too

II

A little while later: The Unmaking of a Poet

Throughout my teenage years, reading and writing became my biggest escapes. They were a way to make sense of the trauma and loss I had faced, but the more I wrote, the more I doubted myself. Did what I had to say really matter?
I still didn't feel like I belonged anywhere or that anyone truly wanted to listen to me. As a dyslexic, sometimes words unfurl, blur and move as I try to read them; the lines escaping me as my eye moves along. That's how this period of my life felt. I was unravelled.

Nine

sacrament of healing

They say that holding on to anger is like holding onto a hot poker. They are wrong.

Holding a hot poker hurts you, but my anger hurts them.

I like the anger, it fuels me, keeps me focused. Lets me know what vengeance I want to seek.

The anger means I will never forget.

They say forgiveness is a virtue. They are wrong.

It is a weak vice. I will not be weak again.

To look them in the eye and forgive my abuser is, quite literally the stupidest thing I've ever heard.

If they wanted forgiveness or mercy, they should have tried *maybe not being an abusive piece of shit?*

Why is it *my* job to fix something that *they* broke?

They say when you are slapped, you should turn the other cheek. They are wrong.

When I am slapped, I slap back. Harder.

How many times must you forgive and forget before you realise, *oops, guess they're never going to change?* I'll never know because I do not forgive *or* forget.

"It'll make you feel better... You'll find peace... It's a better way to live."

Wrong. Wrong. **Wrong**.

What makes me feel good is knowing that if I ever see them again, I will start such a ruckus hitherto unseen.

What makes me feel good is knowing they will

Never.

Ever.

Ever

Get my forgiveness. I am at peace with what happened and I am also at peace with the fact that I will hold that anger for as long as I damn well please. The better way to live is not getting screwed over again and again because you trusted the wrong people.

They say holding onto resentment isn't good for you,

I say *why the hell not?*

Ten

Hospitals, Medicine and Recovery

They say:

> *"You can't help someone who won't help themselves."*

And I guess that is true. But *you* gave up on *me*.
I still believe in you.

Eleven

Midnight Prayers

I

Was...

wandering asleep

I think

I don't know

not sure

The night is playing with my mind

laughing at my tired eyes

bringing back old memories that make me shudder

the shadows growl as they cross my ceiling

the air in my lungs is beginning to freeze

time isn't real

maybe I'm not either

if I am dead, do not bring me back

I wish to slip into the darkness

I long for peace

My god, my God, where are you?

I am not asking for anything
I just want to know if you are there
are you?
hello?

Twelve

Petition B for the will to live

Tell me not to slip into the stars.

Tell me to breathe each moment though it hurts- tell me to live each millisecond of pain. Tell me it's worth it. Tell me there's a point to the feeling.

Tell me why it shouldn't be the end. Tell me it won't be like this forever.

Thirteen

I would

And I would have followed you into the valley of the shadow of death.

I would have...

Opened my arms to your knife,

Let you sink the blade in my waiting spine,

A willing participant to your betrayal.

Clothed myself in your red flags, danced around the warning signs you put up.

Happily, turned my blood into water for you to drink. Eagerly cursed the land with a plague against all men and animals to please you.

I would have followed you into Tartarus, I would have let you swallow me completely in your darkness

I would have...

Held hands with you as you burned your home to the ground. Sat and roasted chestnuts in the flickering, fantastic flames. Thrown your papers in as kindle, watched as your deadname turned to Ash.

I would have left home for you, chased through dead streets, deadly alleyways, deserted desserts to find you.

I would have…

Willingly covered myself in boils and hives, rolled over the ground in mourning for your love. Drowned the world in a hailstorm of fire. Covered the sky in darkness for days on end, just so everyone could feel the pain of you leaving me. I would have given you everything you asked and more, ripped my heart out and eaten it raw, chewed through my own veins, swallowed myself whole, on my knees before you.

All I needed was your love in return. Was that too much to ask for?

Perhaps so. Now I look back and I don't see love: I see obsession and indifference. I thought I was in love when I was really just slowly going insane. Yet, your name still leaves my lips, eager to summon you. My therapist said I'm repeating abusive patterns that I grew up with, that I am accepting less than I deserve because I don't believe I deserve love at all. What does she know? If you came back I would do it again.

For you

I would…

Fourteen

A Visitor Who Belongs Here

I don't know where the sugar is.

I really don't, I'm not being lazy, I swear.

I thought it would be in the cupboard on top of the kettle.

It was there last time I came back.

Or was that in our old house?

I don't know where the pots and pans are either now I think of it.

I wanted to make some Indomine earlier, but the kitchen is now an unfamiliar maze.

Junior plays the piano? Since when?

I thought he would like the hoverboard, but it's okay I can return it and buy some sheet music.

I guess.

No, not really, I just couldn't sleep at all last night.

The bedsheets felt weird. At home I have this blanket- Well not at *home,* home.

I meant at my apartment I have this… Nevermind

Fifteen

Bitter Forgiveness

Sometimes I describe that night,
the one we shared,
as the worst night of my life.
I remember walking away from the club with my friends tired,
but happy.
I remember you yelling,
violent,
drunk.
I remember how it started: insults,
threats
and then
racism.
The racism was not surprising at all
I remember thinking that.
I remember thinking that I couldn't let you see how much the words hurt me.
I remember putting on a tough front.

I remember the fear I felt, feeling like *I might die tonight.*

I remember getting back to my friends' dorm in the early hours of the morning. I remember lying on the bed and immediately bursting into tears.

I remember calling my mum, I remember how scared she was down the phone, her saying I'm *lucky to be alive, and lucky to be one piece.*

I remember when you said to me:

"Call the police you think they'll believe? A nigger or me?"

and my moment of doubt, would the police believe me? Or would they believe the white man in front of me?

The flashbacks come less often now.

I went out with my friends, and I didn't feel my chest tighten on the way home.

I stopped double-checking every corner and alley.

A guy wolf-whistled at us, and I didn't flinch,

We laughed and told him *he looked pretty too.*

My mum said it's because I'm moving on.

I think I might forgive you, one day.

But today, I forgive me.

I forgive myself for the shame and anger I felt about that night.

I forgive myself for the blame and guilt I put on my own shoulders,

but most importantly,

I forgive myself for feeling like I needed forgiveness in the first place.

Sixteen

Truer Love

I once met a girl who loved with all her heart,

Who loved readily and boldly no matter what.

A girl who wasn't afraid to yell her trauma because she had spent far too long being silenced.

A girl who challenged me in the best and worst ways.

I once met a girl who prayed in Shakespeare's sonnets, begging Hermes to remove her anxiety.

A girl who was so sad you could paint her life in blue, or you could paint it as she told it, in all the shades of the rainbow.

A girl who even when she was down had the most beautiful hues you had ever seen, not because her illness was beautiful-but because she was.

I once fell in love with a girl who rightly loved dogs more than humans.

Who found joy in printing museums and long, old books written by dead, old men.

Who never tired of fighting for those she loved.

And I didn't know a soul could fit mine so well.

Seventeen

Apologies in Advance

How can I settle?

No, if my hopes will
not see. I already am
so tired?
Woe,
I settle? No,
if
my
hopes will
not run, I settle?
No, if my hopes
will not
see. I already
"Diss"- with
no matter. This
moment is
good.

If
them.
I feel
a fear, that I'm
already "Diss"- with
no illusioned with no
illusioned with
no
illusioned without
being appointed.
How deluded with no
illusioned with no illusioned without
being appointed. How can I settle? No, if my hopes will not
see. I already
sorry,
for all push they dream
like.
Should I shall push
they
dream like.
I feel a fear,
that

Eighteen

So Obsessive-Compulsive Disorder

You're like so OCD, that's so funny.

So am I

But I think maybe...

I think, maybe, we are talking but slightly different things.

You see *I've* got Obsessive-Compulsive Disorder. It's an anxiety disorder in which people have recurring, unwanted thoughts, ideas or sensations (obsessions) that make them feel driven to do something repetitively (compulsions).

When the GP first diagnosed me, I told her *you are wrong*.

You're wrong.

I don't care about clean things or dirty things. I don't enjoy cleaning things at all. In fact, I hate cleanings this. I can't clean things. I hate this, ~~I hate this conversation~~.

But she said I didn't understand. Because she said I have- I have obsessions.

I have a lot of obsessions.

I'm obsessed with bathrooms and whether they are clean or not.

The obsession isn't rational it's just that I look at things and my OCD knows instantly they aren't clean.

Take for example white walls, I hate white walls and I can't explain it.

Anything with white walls immediately makes me uncomfortable, especially bathroom walls or white tiles.

It's just unclean.

I know it makes no sense. I know in theory white should be the cleanest colour, right? Dirt shows up on white surfaces.

But it's not clean okay. It's just not.

When I'm in a place that triggers my OCD, I get anxious- but it's not normal anxiety.

You know the kind of anxiety you get before an exam, feeling like you left your phone in your pocket with the volume on loud even though you handed it in an hour ago.

But it's not that kind of anxiety, its physical, it consumes **everything**

it's like my chest starts to tighten and I start breathing weirdly ~~but I try to hide it if other people notice it gets worse.~~ It's like my body starts to twitch, and I can't control it -I can't control anything.

My mind starts to focus only on the fact that this place is unclean.

It's unclean.

It's unclean.

I can't focus on anything else.

Only the fact that I have to leave, I have to leave because if you stay in a dirty room you become dirty and then I will be dirty and I cannot-

You see, I thought then that I didn't have any compulsions, but I do have a compulsion but the compulsion- you see it's funny- because the compulsion is just to leave, just to leave. To leave anywhere, anything that might be dirty. I can't stay there. I can't risk the contamination, so I have to leave.

Yeah, it's strange.

But *look at the silver lining* you might think. *Don't be so damn pessimistic.*

No one likes being around a depressed person.

Yeah, I know, I'm around me all the time.

So, I need to be more positive, at least my OCD means that my place is always clean right? Wrong. It's like my mum said when I told her I got diagnosed, she said *that makes no sense you can't have OCD your room is a pigsty*. And she's not wrong, my room is always a mess. But it is a clean mess, everything there is clean. I have an area, okay, where dirty things are allowed. It may seem disorganised to you but it's my mess and my OCD says it's all clean, so it is clean. My brain says these things are okay to interact with these things are okay to touch. They are clean.

But OCD gets worse okay, so it's like this.

I love going home, I love seeing my family. But I don't do it often. In fact, I often avoid it. Because to see my family I have to get on the train.

I hate the train.

I hate public transportation. ~~I love what it does for our economy, and for the environment.~~ It's disgusting. And I hate it.

When I sit on the train I **must** sit backwards facing. I **must** see where we are coming from. I **must** be at a table. I **must** be near a luggage rack but far, far away from the toilet.

~~It's not clean near the toilets the air is-~~

So I must be at a table, but that starts another set of problems. Because the chairs are gross, they look dirty.

So, I have to bring a blanket, a clean blanket that's marked. One side of the blanket can touch clean things the other side cannot. So, I put the dirty side on the chair and then I can sit down. But the windowsills they are dirty too, and they are dusty. And the table, God knows what people put on that table. I don't know what's been on the table. So, I have wipes, Dettol wipes. They kill the flu virus you know. So, I wipe everything down. The table. The window. The armrests. And I hate it because when you clean dirty things you become dirty. So, I have anti-bac gel and I clean my hands thoroughly.

And then do you know what I do for the rest of the train ride? I sit, maybe 30 seconds away from a panic attack the entire journey. Because despite all my work, the train is still not clean. I'm sitting there, and I am just a ball of anxiety and worry contemplating how dirty everything and it's a 5-hour train journey so it's not a fun ride.

OCD is not a fun ride. Did you know that there are different types of OCD?

Oh yeah, there's sexuality OCD, where you doubt your sexuality. And it affects people of all sexualities. But it's not the usual doubts, oh no its *intrusive thoughts*.

So, it's *way* more fun.

There is paedophilia OCD. Where someone's brains keeps telling them that they are a paedophile despite there being no evidence of it. It often happens after they have been abused, or someone they trusted is revealed to be an abuser.

Suddenly their brain doesn't know how to cope

if that person was an abuser, can anyone be?

Is everyone?

Am I?

What if I am?

What if I am evil?

I joined an OCD support group over the summer and there was a girl whose Dad abused her Mum. Her Dad also eventually killed her Mum. So her OCD was based on the fear of becoming an abusive partner. There was no evidence of this. She was 17. But that didn't matter. Her OCD kept telling her, *what if you are just like him?* It kept playing in her mind.

Her compulsion was to hurt herself.

To stay at home.

To turn the light on and of 50 times, throw her keys and catch them 15 times, tap the door thrice all to make sure she would not be a threat to others when she had to go outside to buy food. She never let people in, just in case, she became abusive. 17 and she had never held hands with anyone. She had never had a best friend. She had never gone on a date. She had no one to tell her secrets to. Only me and our OCD support group. I wish I could tell you that she's doing better now. But she's not.

When it destroys OCD it does so with such devastation. And like all mental illnesses, it is so hard to fight, because how do you argue with your own mind? You know all your fears and insecurities, and you know all the best ways to hurt yourself. And the thoughts feel unstoppable. That's why they are called intrusive thoughts- because you can't control them. Then it becomes almost a game and feels so utterly helpless.

Because when you can't face getting up in the morning and showering because the white tiles will make you have an anxiety attack,

when you can't drink milk with your cereal because milk is white and that is bad,

when you must wash your clothes 3 times to make them clean,

when you cannot function like a normal person,

when you're **every thought** is consumed by these obsessions,

you begin to think*: if this is how my life will be, I don't know if I want to live anymore. I think I want to- I think I want to...*

not live.

Wherever you go, there you are.

Wherever you go, there you are.

It's one of my favourite quotes

And I believe it's true because what I would do, *oh God what I would do* to be able to go somewhere without me. Without my thoughts.

It's not entirely hopeless, it just feels that way sometimes. I'm in therapy. I'm getting better. I have an amazing support system.

I don't tell you all this to worry you. It may not feel like I am moving in the right direction, but I am not going backwards at least. Not anymore.

But, anyway enough about me.

I could talk about this for hours.

I want to hear from you.

Tell me about you,

you said that the way that those pens are arranged really annoyed you because you're *so OCD.*

III

After OCD: Hunger

After being diagnosed with OCD, a lot of my childhood began to make sense. I was furious I had wasted so much time.
I had been postponing my own healing all my life, and now I had no choice.
I knew it would be hard but I decided to stop wasting my time. And thankfully, letting go of a lot of dark feelings left space for new ones to enter my life. I suddenly realised that there was so much more to life. And I wanted to have it all.

Nineteen

On Ayo's couch

What if I am the one holding me back?

What do you mean?

I don't miss the days of binging and purging, ~~binging and purging, binging and purging, binging and purging, binging and purging, binging and purging~~ starving myself when I was bad and vomiting up treat food when I was good. I don't *miss* hurting my body like that. Not exactly.

Hmm

I don't want to hurt my body.
I don't want to hate my body.
I *want* to love my body.

But...?

But I can't help longing to be thin.

What can you do to combat that?

All I can do is be myself and desperately hope the world will make room for me.

If the world doesn't, I guess I'll have to make my own space.

~~*But I can't help wondering if I could find my 'real' self through losing a couple pounds more of current myself...*~~

Twenty

Petition C for the will to live

Tell me that I stop writing sad poems.

I want to move on
I want to be happy
I want to find joy in the little things
I want to laugh at a stupidly big orange
I want to stop taking pills to make sure I am alive when I go to pick up *more pills*
Tell me that happens
Please
Tell me anything
I am sorry
Please
Just talk to me

Twenty-One

Hunger

I have this feeling that comes around every now and then.

I didn't recognise it at first, it was so unfamiliar.

It is a deep, guttural feeling, like thunder in my belly.

It makes me think, maybe I don't have to turn back now.

Since I have *cravings* now.

I want love, joy, laughter, true connection, companionship, trust, knowledge, good food and even better sex.

Sometimes the feeling is so deep, I feel like I *deserve* all these things I want.

I guess I didn't remember how it good felt to eat without keeping an eye on the scale.

Maybe, just maybe, you were wrong Mum.

Maybe some things do taste better than skinny feels.

Twenty-Two

Places, Spaces and Belonging

They say:

> *"You can never go home again."*

And I hope that is not true.

I struggle to be truly me in places without you.

Twenty-Three

reconciliation (confessions of sins)

Forgive me, Father, for I have sinned

I have dreamed of doing more than just living to serve you.
I have loved the sin and the loved sinner
I did not hate her touch or her tender kiss.

I know I said last time was the last time, I know I made a covenant with You, but I know my lies never convinced You. I believed them, but not for long.

Forgive me, Father, for I have sinned

I have welcomed those You cast out with open arms, cradling them to my breast. I fed them with the fruits of Your labour, I have bathed them in love. I did not shun them; I elevated their voice when they spoke out against You.

I see the suffering You allow, and I do not understand it.

Forgive me, Father, for I have sinned

Forgive me, for I no longer care what you think.

I have stopped lying to myself

The real sin is you creating Her and cursing me to live without Her touch. In creating love, sex, commitment, joy and expecting me to abstain from them all.

In creating a world and abandoning it to our will, leaving it to rot in the hands of your favourite creation.

Forgive me, or don't.

It matters not to me.

Perhaps it is not whether *you* will forgive *me* in the end, but whether

We will forgive *you.*

Twenty-Four

race

Everyone wants me to talk about race

And I don't know how to say I don't want to anymore
It's your turn.
I want to sit and read,
I want to kiss pretty girls,
and sing in the rain
But I don't have that luxury
I am an *unpaid, untrained, unqualified* ambassador for my people
My existence is a political debate
My whispers are screams into the abyss
Sometimes I touch my throat
just to check that Ursula hasn't stolen my voice
because when I speak,
no one seems to hear me

Twenty-Five

Quaran-Teen

There are 4 teenagers in my house right now, and I am not one of them.

There are also 6 children that I am avoiding.

I am not a teen or a child.

I haven't been for a while.

Strange.

Scary.

We had a socially distanced BBQ, (don't worry BoJo said its alright) restrictions lifted, time to drink prosecco and leave the offspring in the garden.

A little kid came up to me and asked where their brother was.

I looked up from my screen and shrugged:

Beere lowo ikan ninu awon omo yen.

They said they had already.

I had come this far out of the party to avoid this type of interaction.

Go ask an adult then.

The kid looked at me puzzled:

You are *an adult, Aunty.*

Aunty?

We stared at each other in silence.

When did I become an Aunty?

When did I become too old for kids to call me by my first name?

Okay, let's look for your brother.

We walked hand in hand up to the kids running around playing Roblox, or maybe it was Minecraft.

Loud bass from the living room speakers to match the intensity of gameplay, and the loud shouts of glee.

Guys, can we turn that down just a little, I-

Silence, immediately.

A chorus of *"Yes Aunty!"* follows, and I realise: *Yes,* I *am* an Aunty now.

That's how they see me because that's who I am now.

Too old to be a teen, too young to be one of the parents.

Okay, maybe I can get used to this.

I find the kid's brother in no time, in the office playing Fortnite with the preteens.

I deliberately call it *'The fortnights'* and laugh at them cringing at how uncool I am.

Its been a while, so I hadn't noticed, but here I was: An Aunty.

Now I ate before the kids, got the biggest meat with my Jollof, chatted with grandma about the youth and how wayward they were becoming.

A teen tried to dobale for me and I laughed and pulled them up.

That's all it took, one quarantine and I went from an awkward teen to an aunty.

Twenty-Six

Dear 13-year-old me

I wish I had something profound to say, but we both know that's not our forte.

Or maybe it is. Time will tell. That's what I want to tell you; that time will be your greatest love and your worst enemy. It will grow your boys up so fast- sooner than you can blink- then it'll steal your innocence and make you weep. And if you think you've cried before, let's just say 17 is a banner year for new records of all kinds.

Life is going to suck, a lot, and I really wish it wouldn't- but I also don't know who we would be without the suffering. That doesn't mean we are thankful for it. It just means that things could have been different (and you'll spend many hours wishing and imagining that they were) but different doesn't always mean better.

One thing I have learnt though is the importance of unsticking yourself. We tend to get stuck on things, try your best not to. Things get bad, but there are always beautiful moments ahead. It might get better, it might get worse- but if you stop living, you will never get to see how it ends.

And you know how much we hate a cliff-hanger.

Most of all, I wish I could give you a hug.

I spent most of today so angry, at you, me, the world, fate: everything. But now I am sitting on soft grass, eating crab legs, and watching our goldfish swim around our pond. The sky is beautiful, the air is clean, our family is safe. Life gets good. It doesn't always feel that way, sometimes it feels awful, but there is some light in the dark- and it is ours to claim or not.

Looking back at your letters to me, you only want one thing really: for us to be happy. Am I happy?

Almost,

maybe,

someday.

You've got this. You'll see me sooner than you think.

Lots of Love,

Future you.

(P.S. You're gay.

I know, it's crazy, look it up.)

Twenty-Seven

Sappho told me to

For what its worth, I think you should kiss the girl

Kiss the girl
Seriously, kiss the girl
Time is short
She can say no
That's okay
You will fracture but you will not break
Nothing can break you now
Do you hear me?
Nothing
Look her in the eye
Reach out
And just say it
Say what you feel,
That slender Aphrodite has overcome you,
with longing for a girl
Hold her hand

When they jeer, hold it tighter
Be soft with her, and let her touch your soul
Life is less without love,
whatever kind of love you desire
and you are enough, by yourself-
don't let a handsome girl make you forget
But she *is* handsome
So, for what it is worth, I think you should kiss the girl

Acknowledgements

I don't really know what to say here, this anthology doesn't feel finished and in some ways, it is not. The poems I wrote tell of my journey through life, from when I could first put pen to paper to the tentative adulthood I now find myself in. So, it doesn't quite yet feel finished, because I don't know how it will all end. I just hope its epic.

Of course, I must say a huge thanks to Lucile Cliffton: may she always rest in power.

Many thanks to ìyá mi even if I could turn every word on these pages into a thank you, it would still not be enough acknowledgement for all you have done for me. Despite that, every word I write now and until the end of time is dedicated to you and your endless love.

Thanks to David, for showing my mother that love can be patient, kind, forgiving and most importantly: good.

To my siblings: Stop using my Netflix profile. Seriously, you have your own, which I paid for! Also thank you for being my best friends, my worst enemies, my confidants, my bullies, my haven, my hell, my brothers, my sisters and everything in between.

Thanks to Vicky for all the laughs, jokes, and roasts that made me who I am. There's no one like you. Thanks to Emily for seeing my soul and helping me translate it. Thanks to Bolu: You are an artistic genius and your vision is insane, thank you for lending your talents to the cover of my anthology.

Thanks to Miss Childs, Mr Austin, Dr A C Machacek for seeing something special in me.

Finally, thank you dear reader: I hope it isn't presumptuous of me to say that I hope this book helped you in some way. I am still striving to be the woman 13-year-old me needed to help her, and this work is only part of that journey. Thank you for helping that journey along, and please remember that you have your own story to tell and it is worth sharing! Don't let **anyone** take that away.

I spent far too long considering if I should publish this anthology, and waiting for a sign. So, if you are like me and hesitating over something, desperate for a sign to do that thing that makes your heart soar and your mind sparkle: This is it. **This is your sign.**

You've got this, *truly,* you do.

About the Author

Luwa Adebanjo is a poet, writer, actress and theatre-maker originally from Nigeria. As an immigrant coming to the UK, she struggled to fit in. Luwa began reading at a young age and began writing her own works at the age of 11. Since then it has been her dream to be an acclaimed author.

She loves Mexican food, writing, binge-watching Netflix and board games in equal amounts. She hates celery, essay deadlines and having to write her own 3rd person bio in increasing amounts respectively.

You can read more of Luwa's work on her Medium page, where she writes about social issues, news, life and else whatever is on her mind.

You can connect with me on:

https://linktr.ee/LuwaAdebanjo

https://www.facebook.com/LuwaAdebanjoCreates

https://luwaadebanjo.medium.com

Also by Luwa Adebanjo

We're Here for Laura

We're Here for Laura is a comedy play about four horrible friends.

Laura Penbrooke is dead. Her four closest friends can't wait to milk her tragedy for all it is worth. No one more so than Mitton, Laura's bestie, who arranges a simple dinner to remember her BBFL. So what if her camera crew just happens to be around filming her reality show? That's just showbiz, baby! And if Paul wants to plug their one-person opera, that's a great distraction from Carol's monotone critiques and Alex's latest lifestyle change. Things may get a little bit out of hand, but as long as they all remember that they are 'here for Laura', the night is bound to be a quiet, respectful night of remembrance. Right? Right. Lights, camera, action!

.

Printed in Great Britain
by Amazon

56661464R00047